Collins

easy learning

Multiplication and division

Ages 5–7

8 ÷ 2 = 4

Peter Clarke

How to use this book

- Find a quiet, comfortable place to work, away from other distractions.
- Ask your child what multiplication and division topic they are doing at school and choose an appropriate topic.
- Tackle one topic at a time.
- Help with reading the instructions where necessary and ensure that your child understands what they are required to do.
- Help and encourage your child to check their own answers as they complete each activity.
- Discuss with your child what they have learnt.
- Let your child return to their favourite pages once they have been completed, to play the games and talk about the activities.
- Reward your child with plenty of praise and encouragement.

Special features

- **Yellow boxes:** Introduce and outline the key multiplication or division ideas.
- **Example boxes:** Show how to do the activity.
- **Orange shaded boxes:** Offer advice to parents on how to consolidate your child's understanding.
- **Games:** Some of the topics include a game, which reinforces the multiplication or division topic. Some of these games require a spinner. This is easily made using a pencil, a paperclip and the circle printed on each games page. Place the pencil and paperclip at the centre of the circle and then flick the paperclip to see where it lands.

Published by Collins
An imprint of HarperCollins*Publishers*
1 London Bridge Street
London SE1 9GF

Browse the complete Collins catalogue at www.collins.co.uk

© HarperCollins*Publishers* 2011
This edition © HarperCollins*Publishers* 2015

13

ISBN-13 978-0-00-813434-1

MIX
Paper from responsible sources
FSC™ C007454

The author asserts his moral right to be identified as the author of this work.

The author wishes to thank Brian Molyneaux for his valuable contribution to this publication.

All rights reserved. No part of this publication may be reproduced, stored in a retrieval system, or transmitted, in any form or by any means, electronic, mechanical, photocopying, recording or otherwise, without the prior permission of Collins.

British Library Cataloguing in Publication Data

A Catalogue record for this publication is available from the British Library

Written by Peter Clarke
Page design by G Brasnett, Cambridge and Jouve
Illustrated by Katy Dynes, Rachel Annie Bridgen and Andy Tudor
Cover design by Sarah Duxbury and Paul Oates
Cover illustration by Kathy Baxendale
Project managed by Chantal Peacock and Sonia Dawkins

Printed in Great Britain by Bell and Bain Ltd, Glasgow

Contents

How to use this book	2
Understanding multiplication	4
Understanding division	6
Combining groups of 2, 5 and 10	8
Arranging into equal groups of 2, 5 and 10	10
Practising division	12
Counting in steps of 1, 2, 5 and 10	14
Doubling and halving	16
2 times table	18
Division facts related to the 2 times table	20
5 times table	22
Division facts related to the 5 times table	24
10 times table	26
Division facts related to the 10 times table	28
Linking multiplication and division	30
Answers	32

Understanding multiplication

We can think of multiplication as counting on in equal groups – **repeated addition**.

Example
5 + 5 + 5 + 5 = 20
4 lots of 5 is 20.
4 × 5 = 20

1 Draw lines to match each sandcastle with one addition and one multiplication number sentence.

2 + 2 + 2 = 6

5 × 4 = 20

4 + 4 + 4 + 4 + 4 = 20

3 × 2 = 6

4 × 6 = 24

6 + 6 + 6 + 6 = 24

2 How many stars on each set of coloured flags? Write an addition and multiplication number sentence for each set of flags.

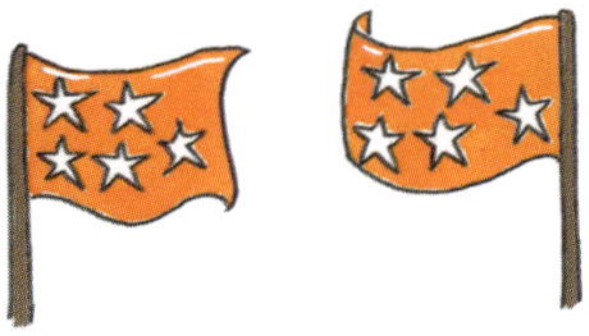

4

We can also think of multiplication as a picture called an **array**.

$4 \times 5 = 20$ gives the same answer as $5 \times 4 = 20$.

Example

$5 + 5 + 5 + 5 = 20$
4 lots of 5 is 20.
$4 \times 5 = 20$

$4 + 4 + 4 + 4 + 4 = 20$
5 lots of 4 is 20.
$5 \times 4 = 20$

3 Draw lines to match each picture to two multiplication number sentences.

$6 \times 3 = 18$

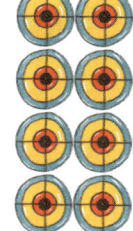

$2 \times 7 = 14$

$2 \times 4 = 8$

$4 \times 2 = 8$

$3 \times 6 = 18$

$7 \times 2 = 14$

4 Write two multiplication number sentences for each picture.

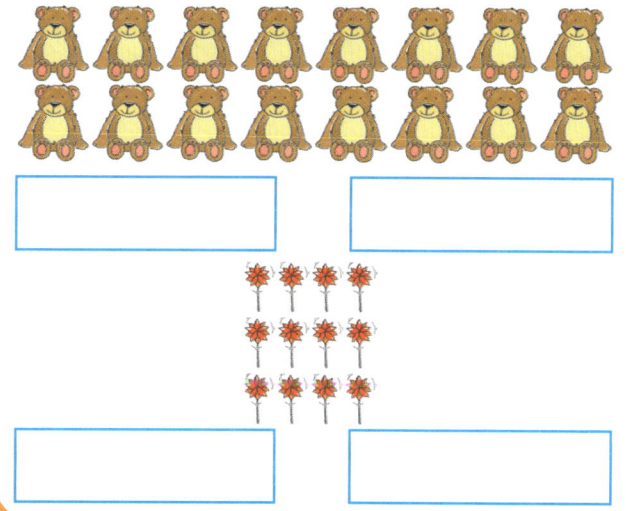

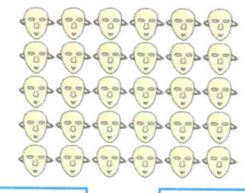

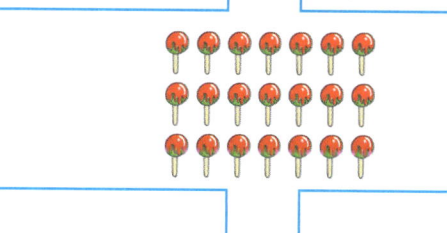

At this stage, your child is beginning to learn about multiplication as repeated addition. An array is a powerful model and image that teaches your child about the commutative law as it applies to multiplication, that is, that 4×5 is the same as 5×4.

Understanding division

We can think of division as **sharing**.

Example Share 12 sweets between 3 plates.
$12 \div 3 = 4$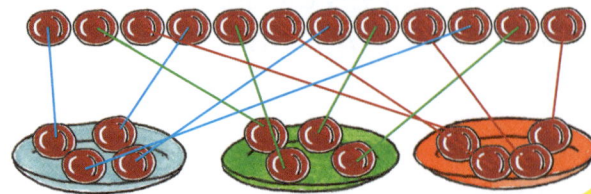

1 Draw lines to match the bags of sweets to the plates and to the division number sentence.

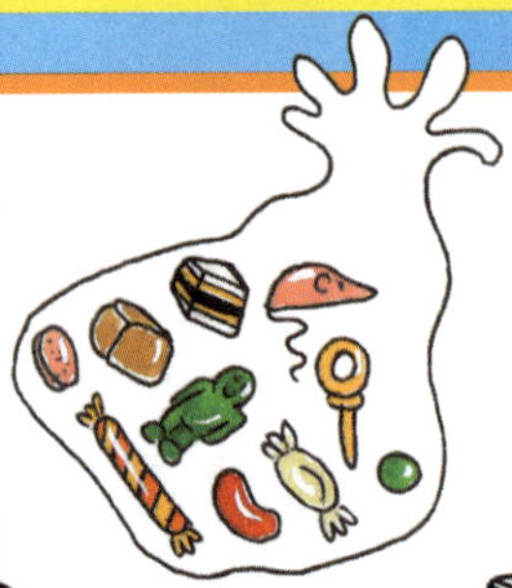

$10 \div 5 = 2$ $6 \div 2 = 3$ $16 \div 4 = 4$

2 Share these sweets. Then write the answer to the division number sentence.

Share 8 sweets between 4.

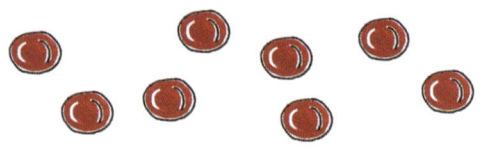

$8 \div 4 = \square$

Share 10 sweets between 2.

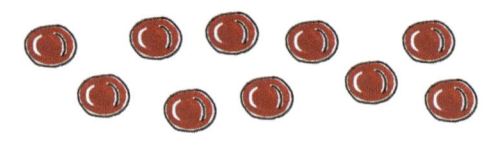

$10 \div 2 = \square$

Share 9 sweets between 3.

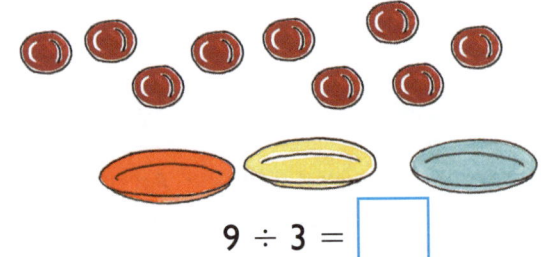

$9 \div 3 = \square$

Share 12 sweets between 6.

$12 \div 6 = \square$

We can also think of division as **grouping**.

Example
How many groups of 5 in 15?
$15 \div 5 = 3$

3 Draw lines to match each plate of chocolates to the division number sentence.

$6 \div 3 = 2$ $20 \div 5 = 4$ $12 \div 4 = 3$

4 Group these sweets. Then write the answer to the division number sentence.

How many groups of 2 in 8?   How many groups of 7 in 14?

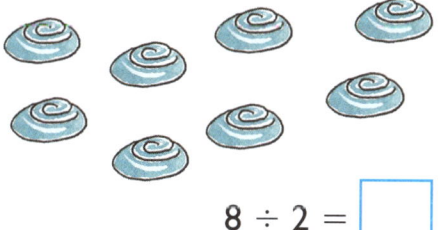 

$8 \div 2 = \square$ $14 \div 7 = \square$

How many groups of 5 in 25?   How many groups of 4 in 24?

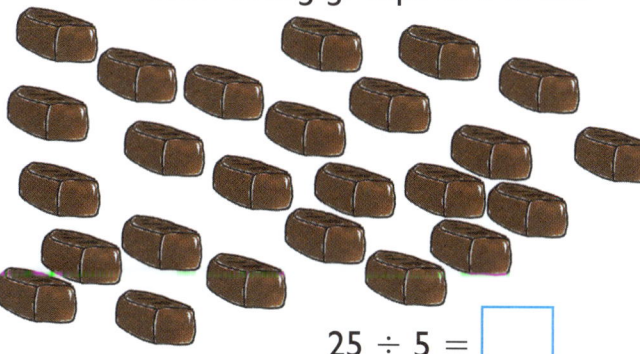 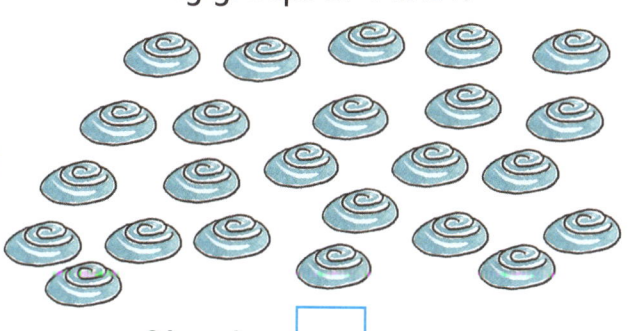

$25 \div 5 = \square$ $24 \div 4 = \square$

There are two ways of looking at division: sharing and grouping. Sharing refers to situations of allocating, where a given amount is shared out equally into a given number of sets. Division as grouping arises from situations where the amount is known as well as the size of each group.

Combining groups of 2, 5 and 10

1 How many wheels on each street?

4 × 2 = ☐

7 × 2 = ☐

3 × 2 = ☐

5 × 2 = ☐

2 How many toes?

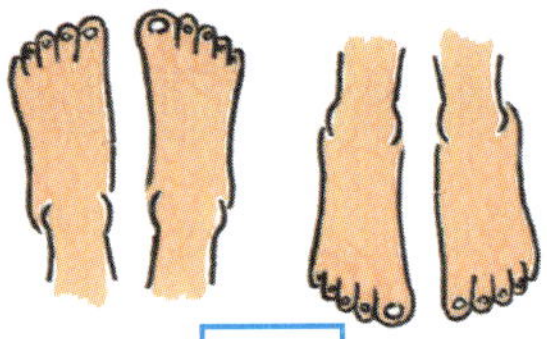

4 × 5 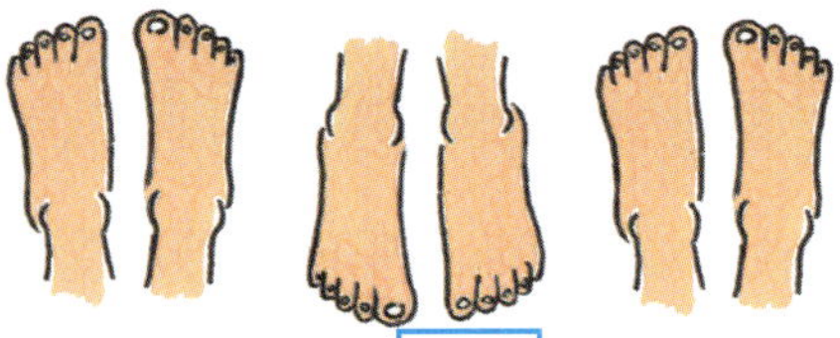 = ☐   6 × 5 = ☐

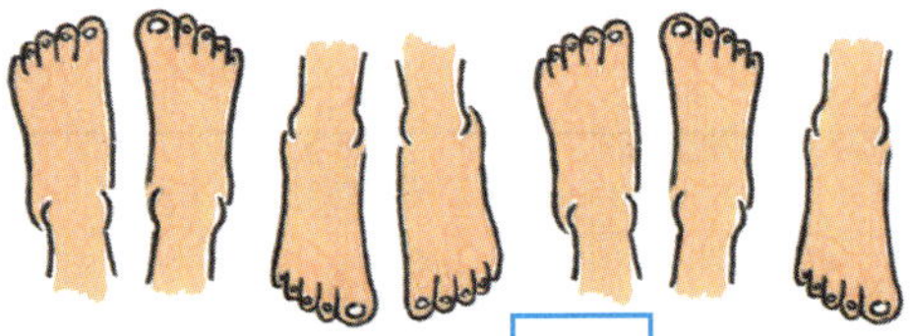

7 × 5 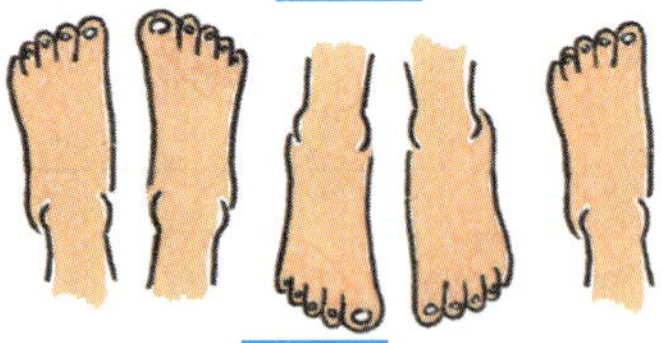 = ☐   5 × 5 = ☐

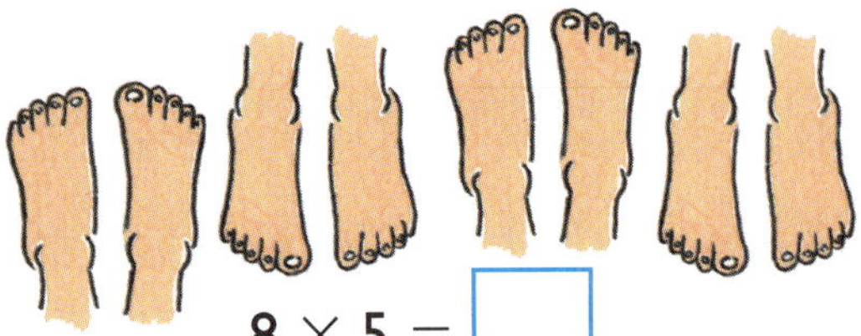

8 × 5 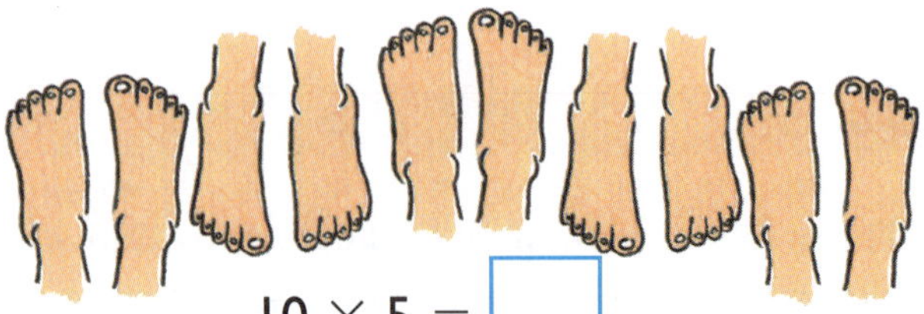 = ☐   10 × 5 = ☐

3 How many cakes?

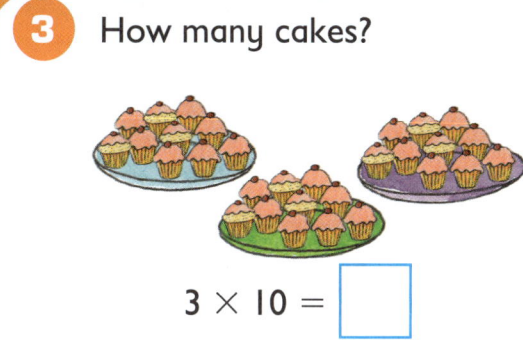

3 × 10 = ☐

6 × 10 = ☐

2 × 10 = ☐

4 × 10 = ☐

7 × 10 = ☐

5 × 10 = ☐

4 Answer these.

How many wheels on 6 bicycles?

6 × 2 = ☐

How many wheels on 9 bicycles?

9 × 2 = ☐

How many toes on 3 feet?

3 × 5 = ☐

How many toes on 12 feet?

12 × 5 = ☐

How many cakes on 8 plates?

8 × 10 = ☐

How many cakes on 11 plates?

11 × 10 = ☐

Coins are a powerful real-life model and image to help your child with combining groups of 2, 5 and 10. Ask your child to count a number of 2p coins, e.g. 4. Then ask: 'So if there are four 2p coins, how much money is this altogether? How do you know?'

Arranging into equal groups of 2, 5 and 10

Example

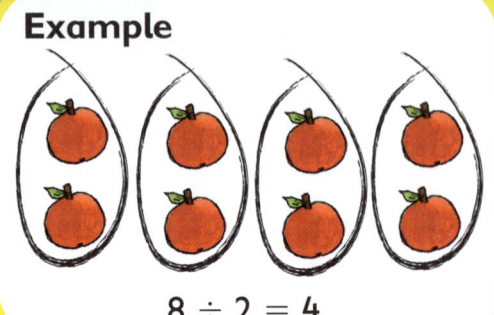

$8 \div 2 = 4$

1 Put the fruit into groups of 2.

$12 \div 2 = \square$

$10 \div 2 = \square$

$16 \div 2 = \square$

2 Put each food into groups of 5.

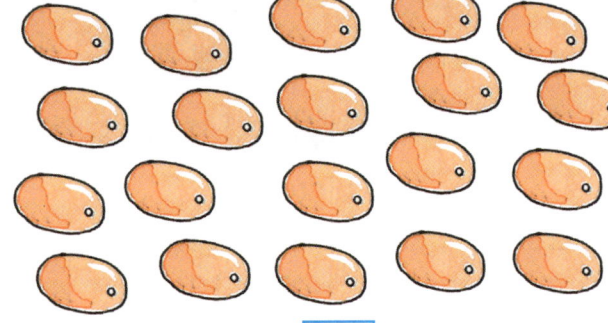

$20 \div 5 = \square$

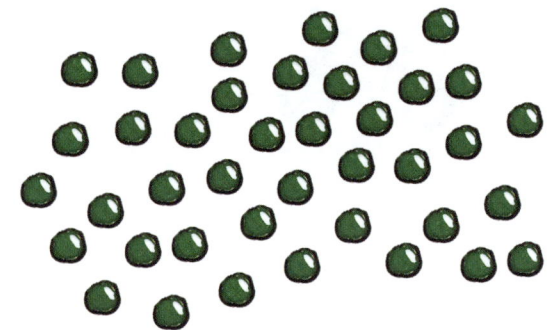

$40 \div 5 = \square$

$15 \div 5 = \square$

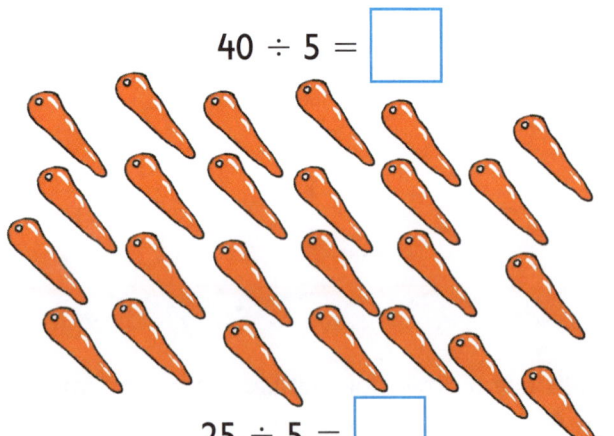

$25 \div 5 = \square$

3 Put each food into groups of 10.

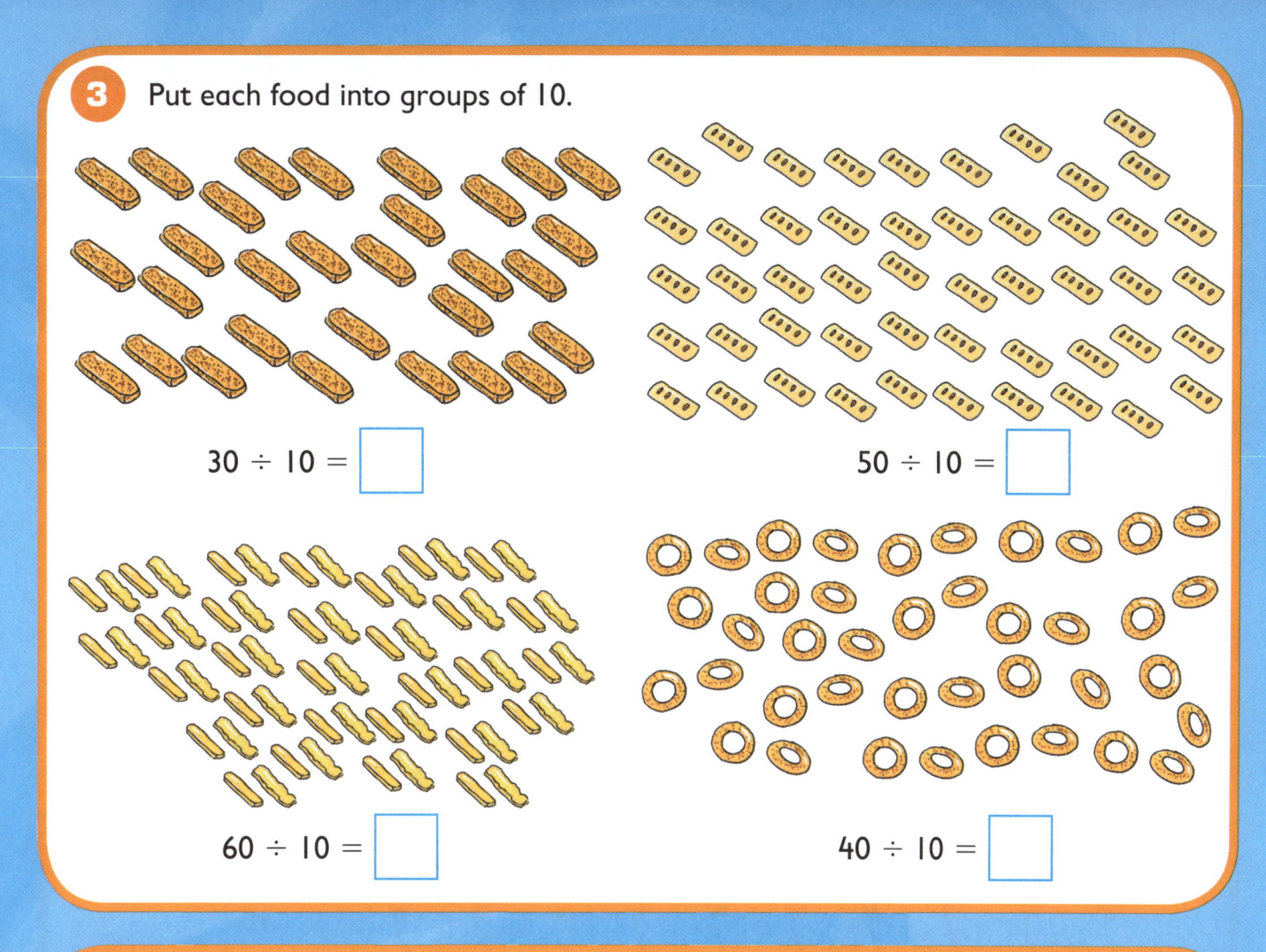

30 ÷ 10 = ☐

50 ÷ 10 = ☐

60 ÷ 10 = ☐

40 ÷ 10 = ☐

4 Put food into groups of 2, 5 or 10. Then write the division number sentence.

How many 2s in 14?

☐ ÷ ☐ = ☐

How many 5s in 10?

☐ ÷ ☐ = ☐

How many 5s in 35?

☐ ÷ ☐ = ☐

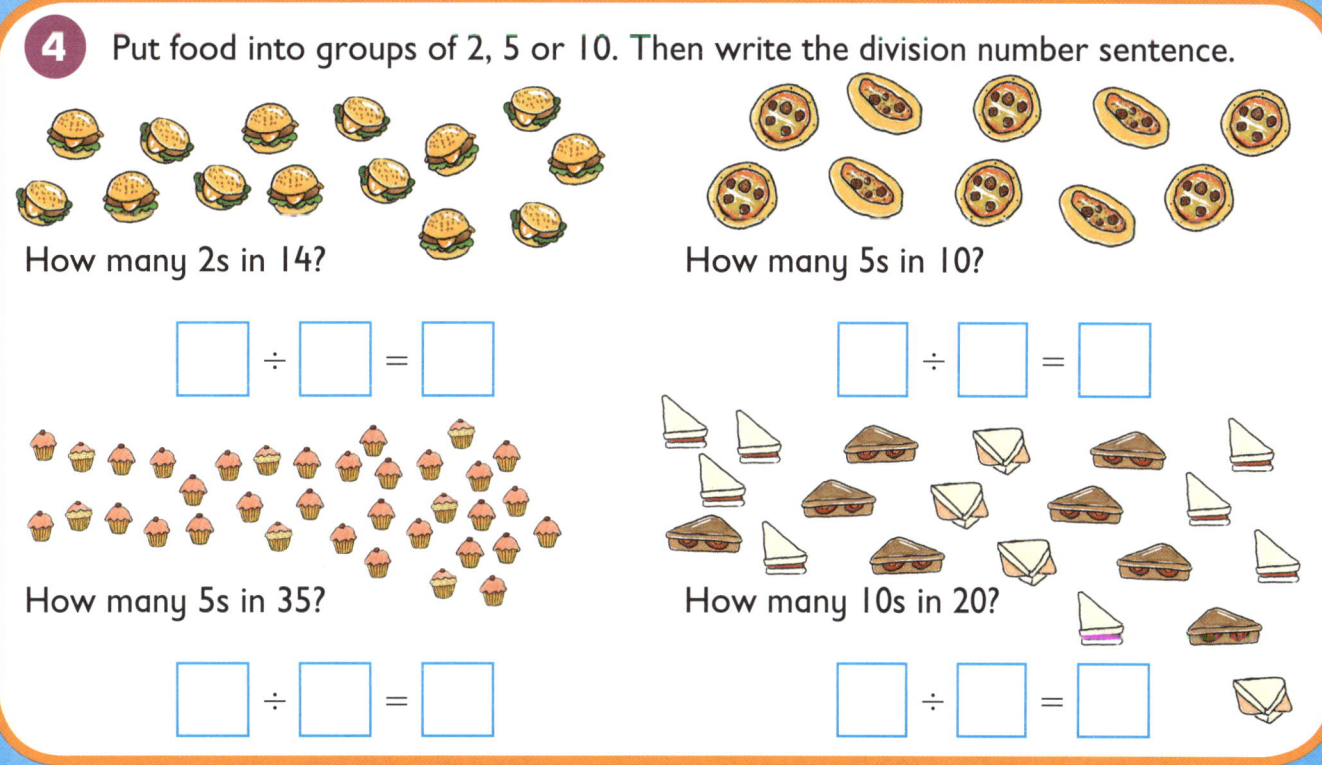

How many 10s in 20?

☐ ÷ ☐ = ☐

Place 2 small plates in front of your child and give them an even number of pasta shells (or similar), e.g. 8. Ask your child to share the 8 pasta shells evenly between the 2 plates. Then ask: 'How many shells are there on each plate?' Repeat for other even numbers to 24.

Practising division

Example 12 ÷ 2 = 6

1 Divide each set of flowers into groups of 2.

8 ÷ 2 = ☐

14 ÷ 2 = ☐

10 ÷ 2 = ☐

18 ÷ 2 = ☐

2 Divide each set of bugs into groups of 5.

10 ÷ 5 = ☐

15 ÷ 5 = ☐

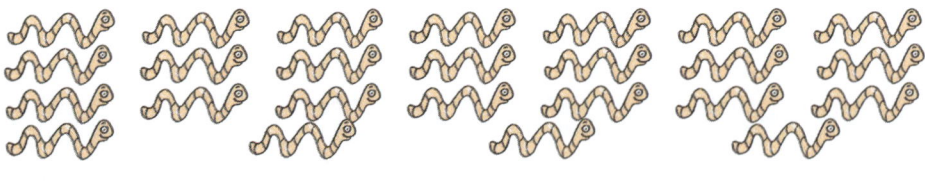

25 ÷ 5 = ☐

30 ÷ 5 = ☐

3 Divide each set of leaves into groups of 10.

$20 \div 10 = \square$

$30 \div 10 = \square$

$40 \div 10 = \square$

4 Put the birds in groups to help you complete each division number sentence.

$12 \div 3 = \square$

$24 \div 4 = \square$

$24 \div 6 = \square$

$28 \div 7 = \square$

In real life, most division situations involve a remainder. This often results in having to round up or down depending on the situation. Ask your child questions that involve divisions with remainders, e.g. A box holds 10 cartons. How many boxes are needed for 23 cartons? The answer is 3 boxes, *not* 2 boxes remainder 3.

Counting in steps of 1, 2, 5 and 10

This number sequence **counts on** in steps of 2.

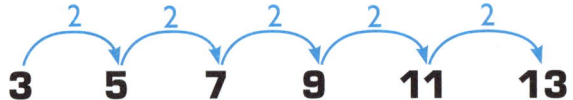

3   5   7   9   11   13

The difference between each number is 2.

1 Write the next 3 numbers in each sequence.

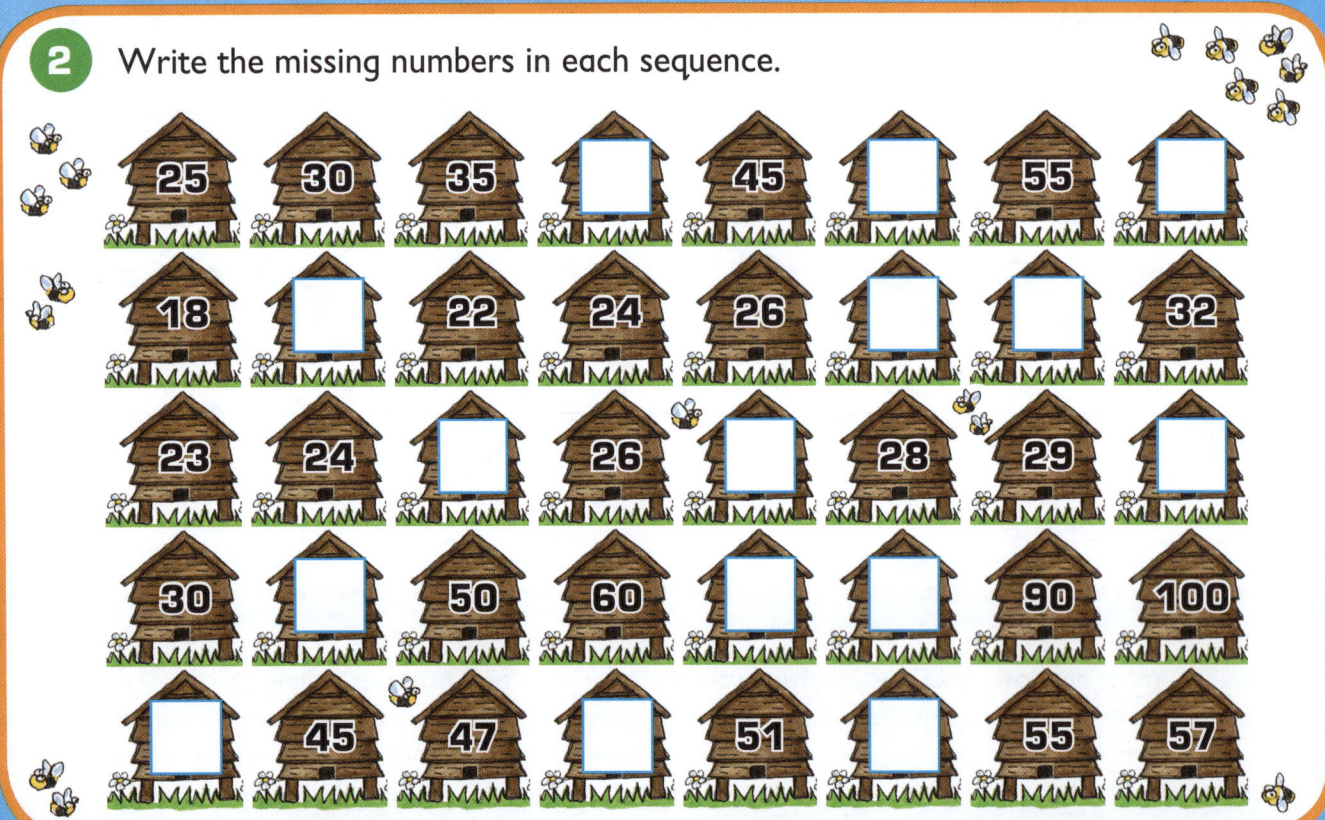

This number sequence **counts back** in steps of 5.

35 30 25 20 15 10

The difference between each number is 5.

3 Write the next 3 numbers in each sequence.

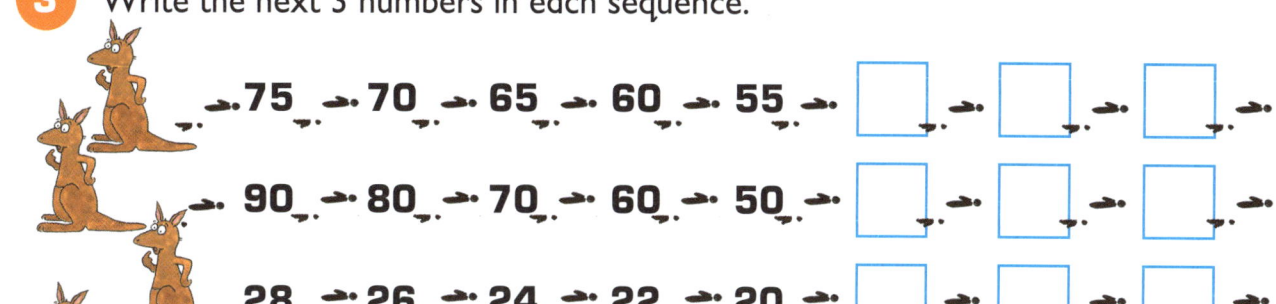

75 → 70 → 65 → 60 → 55 → ☐ → ☐ → ☐
90 → 80 → 70 → 60 → 50 → ☐ → ☐ → ☐
28 → 26 → 24 → 22 → 20 → ☐ → ☐ → ☐
45 → 44 → 43 → 42 → 41 → ☐ → ☐ → ☐
37 → 35 → 33 → 31 → 29 → ☐ → ☐ → ☐

4 Write the missing numbers in each sequence.

34, 33, ☐, 31, ☐, 29, 28, ☐

☐, 90, 85, ☐, 75, ☐, 65, 60

100, ☐, 80, 70, ☐, 50, ☐, 30

65, 60, 55, ☐, 45, ☐, ☐, 30

☐, 84, ☐, 82, 81, 80, ☐, 78

Point to a number, e.g. on a number plate, front door or newspaper. Ask your child to count on or back in ones from that number, e.g. 26. When counting back, ensure that the count does not go below zero. Point to other numbers and ask your child to count on or back in steps of 1, 2, 5 or 10.

Doubling and halving

Doubling is the same as **multiplying by 2**.

　　Double 8 is 16
　　8 + 8 = 16
　　2 × 8 = 16

Halving is the same as **dividing by 2**.

　　Half 20 is 10
　　20 ÷ 2 = 10

1 Double each of these numbers.

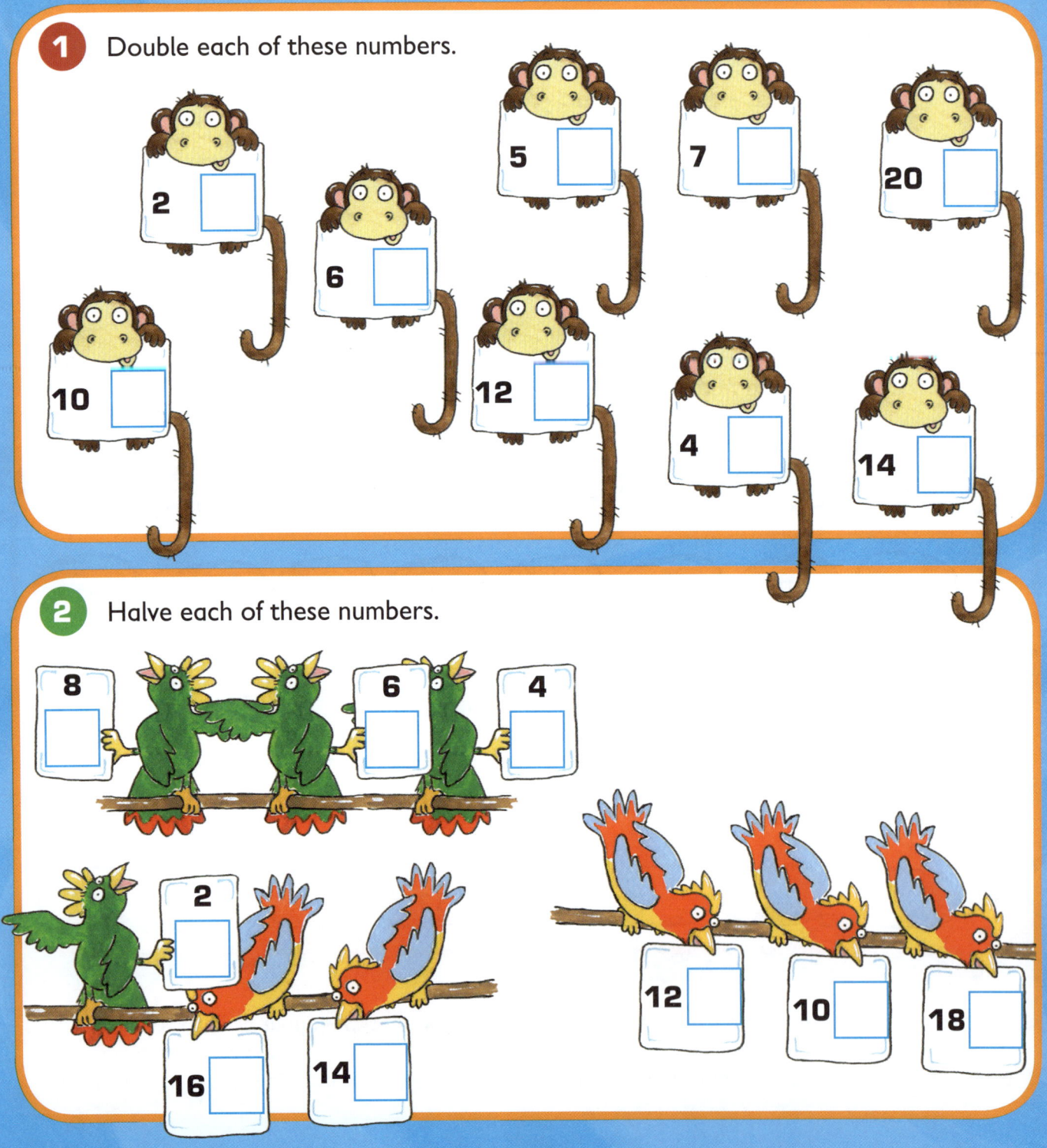

2 Halve each of these numbers.

Game: Two games to play with a partner

You need: paperclip, pencil and 2 different coloured counters

Doubling Game

- Each player takes a counter and places it on 'Start'.
- Take turns to:
  - Spin the spinner and move your counter that number of spaces.
  - Double the number you land on. If your answer is right, move your counter on 1 space. If your answer is wrong, move your counter back 1 space.
- The winner is the first player to pass the 'End'.

Halving Game

- Each player takes a counter and places it on 'Start'.
- Take turns to:
  - Spin the spinner and move your counter that number of spaces.
  - If you land on an odd number, do nothing.
  - If you land on an even number, halve that number. If your answer is right, move your counter on 1 space. If your answer is wrong, move your counter back 1 space.
- The winner is the first player to pass the 'End'.

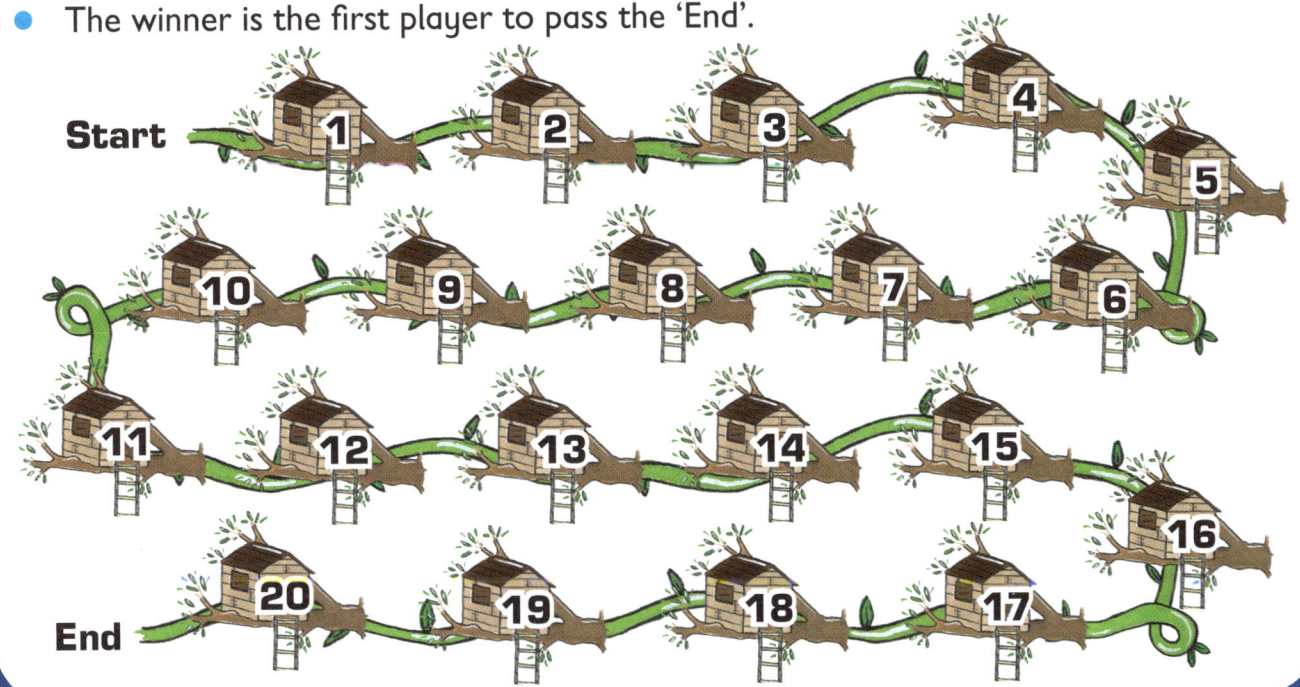

Point to any number on the jungle trail and ask your child to double the number. Use words such as: double, twice, two times. Point to any even number on the jungle trail and ask your child to halve the number. Use words such as: halve, half of, divided by 2.

2 times table

1 Answer these.

4 × 2 = ☐   10 × 2 = ☐   9 × 2 = ☐

2 × 5 = ☐   3 × 2 = ☐   2 × 8 = ☐

7 × 2 = ☐   2 × 6 = ☐   2 × 10 = ☐

2 × 9 = ☐   2 × 2 = ☐   11 × 2 = ☐

2 × 3 = ☐   8 × 2 = ☐   2 × 7 = ☐

2 Write the missing numbers.

6 × ☐ = 12   ☐ × 2 = 2   2 × ☐ = 4

☐ × 2 = 14   2 × ☐ = 10   ☐ × 10 = 20

9 × ☐ = 18   ☐ × 2 = 16   ☐ × 2 = 6

2 × ☐ = 24   ☐ × 2 = 8   ☐ × 2 = 18

☐ × 7 = 14   2 × ☐ = 20   3 × ☐ = 6

Game: 2 times table game

You need: paperclip, pencil and 20 counters

Player 1

| 6 | 12 | 8 | 22 | 10 | 16 | 20 | 24 | 14 | 18 |

Player 2

| 10 | 18 | 24 | 12 | 8 | 14 | 22 | 20 | 16 | 6 |

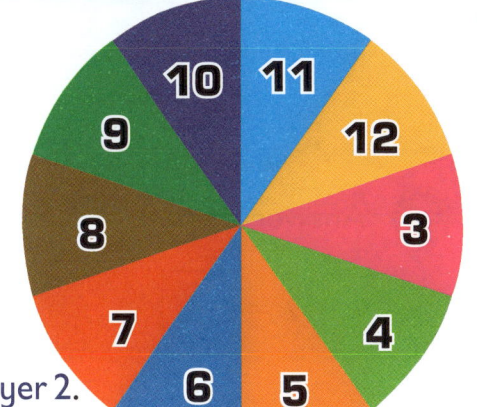

- Before you start decide who is Player 1 and who is Player 2.
- Take turns to:
  – Spin the spinner and say the number.
  – Multiply the number by 2 and say the answer.
  – Place a counter on that number on your line above. If the number already has a counter on it, miss a turn.
- The winner is the first player to cover 8 of their numbers with a counter.

3 For each target, multiply the number hit by 2.

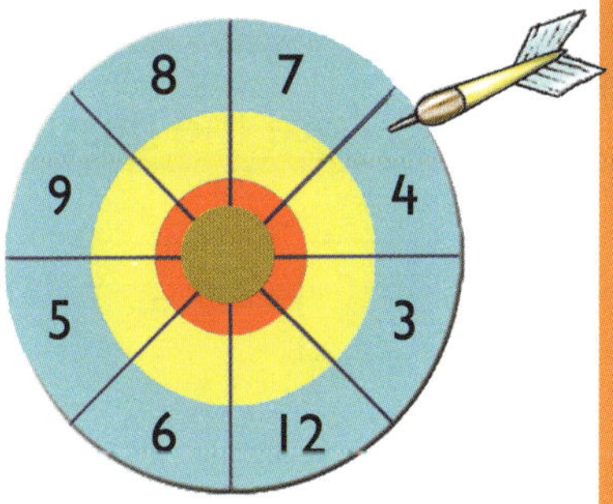

Point to one of the numbers in the game, e.g. 18, and ask your child to tell you the 2 times table fact that has that answer, i.e. 9×2.

19

Division facts related to the 2 times table

1 Answer these.

4 ÷ 2 = ☐   10 ÷ 2 = ☐   14 ÷ 2 = ☐

20 ÷ 2 = ☐   6 ÷ 2 = ☐   18 ÷ 2 = ☐

8 ÷ 2 = ☐   16 ÷ 2 = ☐   22 ÷ 2 = ☐

2 Write the missing numbers.

☐ ÷ 2 = 7   20 ÷ ☐ = 10   ☐ ÷ 2 = 3

18 ÷ ☐ = 9   ☐ ÷ 2 = 5   14 ÷ ☐ = 7

☐ ÷ 2 = 4   ☐ ÷ 2 = 1   ☐ ÷ 2 = 9

☐ ÷ 2 = 11   6 ÷ ☐ = 3   ☐ ÷ 2 = 12

☐ ÷ 2 = 2   ☐ ÷ 2 = 8   8 ÷ ☐ = 4

Game: Dividing by 2 game

You need: paperclip, pencil and 20 counters

Player 1			Player 2		
10 ÷ 2	20 ÷ 2	22 ÷ 2	12 ÷ 2	14 ÷ 2	20 ÷ 2
8 ÷ 2		12 ÷ 2	24 ÷ 2		16 ÷ 2
24 ÷ 2		6 ÷ 2	18 ÷ 2		6 ÷ 2
14 ÷ 2	18 ÷ 2	16 ÷ 2	8 ÷ 2	22 ÷ 2	10 ÷ 2

- Before you start decide who is Player 1 and who is Player 2.
- Take turns to:
    - Spin the spinner and say the number.
    - Look for a 2 times table division fact that equals that number.
    - Place a counter on that fact on your chart above. If the fact already has a counter on it, miss a turn.
- The winner is the first player to cover 8 of their facts with a counter.

3 Divide the numbers hit by 2. Write the answer in the same coloured box as the arrow.

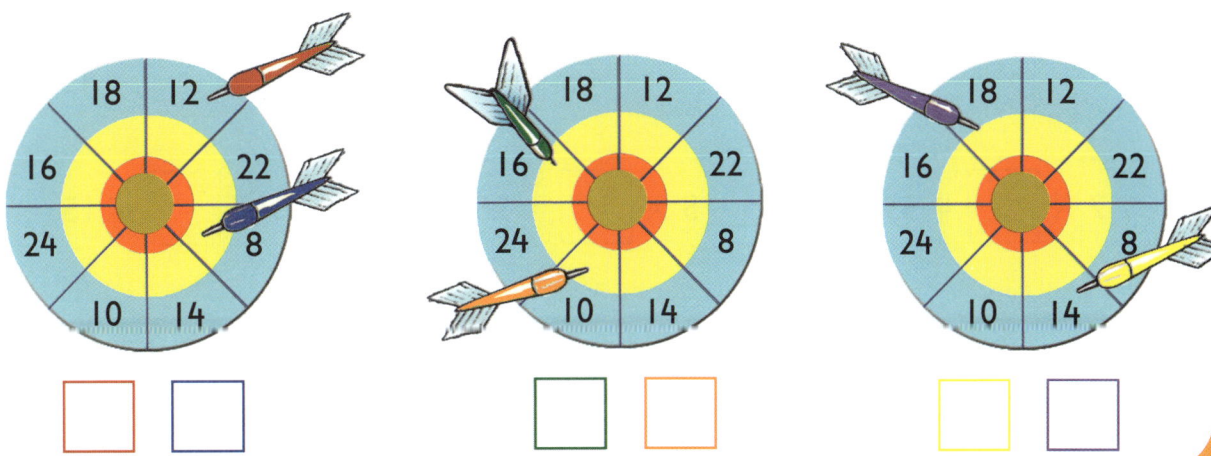

Point to one of the division calculations in the game and ask your child to say the answer. Encourage your child to give quick responses to each question.

5 times table

1 Answer these.

7 × 5 = ☐   5 × 6 = ☐   5 × 10 = ☐

4 × 5 = ☐   10 × 5 = ☐   9 × 5 = ☐

2 × 5 = ☐   11 × 5 = ☐   5 × 8 = ☐

5 × 3 = ☐   8 × 5 = ☐   5 × 7 = ☐

5 × 9 = ☐   6 × 5 = ☐   5 × 5 = ☐

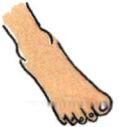   

2 Write the missing numbers.

6 × ☐ = 30   ☐ × 5 = 10   5 × ☐ = 20

☐ × 5 = 5   ☐ × 5 = 40   5 × ☐ = 45

7 × ☐ = 35   5 × ☐ = 25   5 × ☐ = 15

☐ × 2 = 10   ☐ × 5 = 35   ☐ × 5 = 60

9 × ☐ = 45   5 × ☐ = 30   5 × ☐ = 10

Game: 5 times table game

You need: paperclip, pencil and 20 counters

Player 1

| 15 | 40 | 25 | 55 | 50 | 20 | 35 | 60 | 45 | 30 |

Player 2

| 30 | 55 | 45 | 20 | 35 | 60 | 25 | 50 | 15 | 40 |

- Before you start decide who is Player 1 and who is Player 2.
- Take turns to:
  - Spin the spinner and say the number.
  - Multiply the number by 5 and say the answer.
  - Place a counter on that number on your line above. If the number already has a counter on it, miss a turn.
- The winner is the first player to cover 8 of their numbers with a counter.

3 Each number that goes into a function machine is multiplied by 5. Write the numbers that come out of the machines.

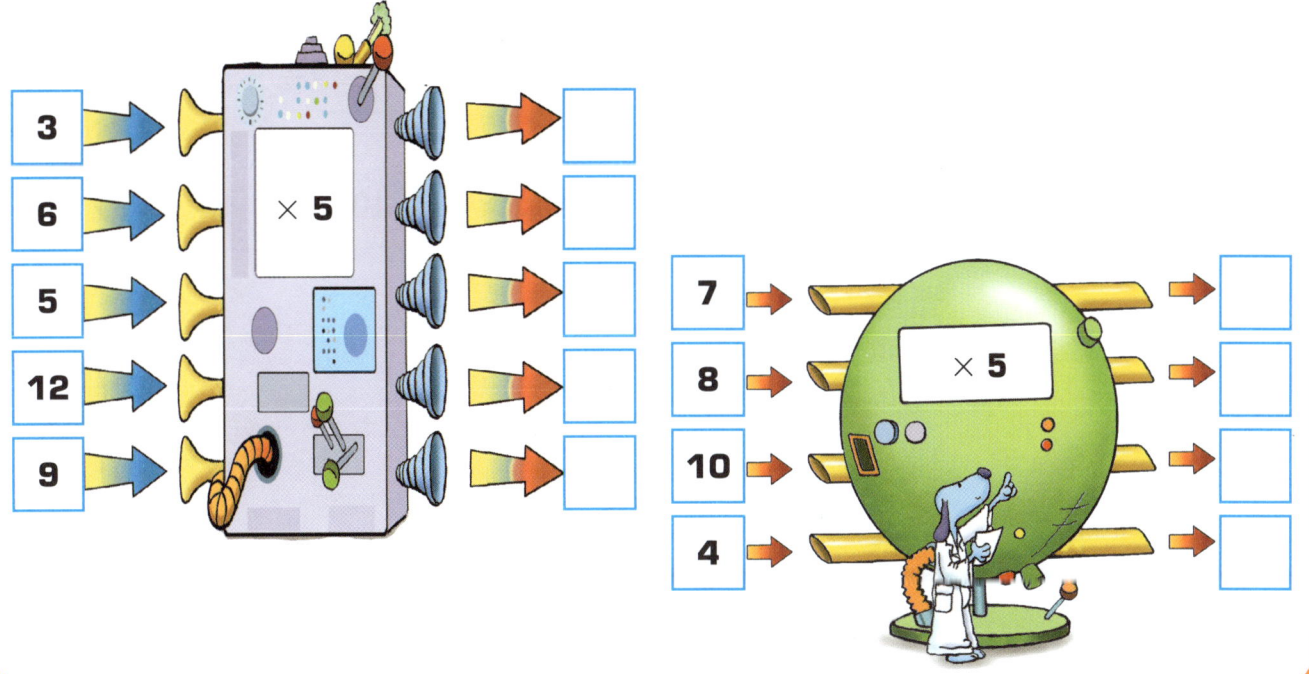

Take turns to spin the spinner. Both players multiply the number by 5. The first player to call out the correct answer wins that round and collects a counter. Play 20 rounds. Who wins the most counters?

Division facts related to the 5 times table

1 Answer these.

20 ÷ 5 = ☐   35 ÷ 5 = ☐   45 ÷ 5 = ☐

30 ÷ 5 = ☐   60 ÷ 5 = ☐   25 ÷ 5 = ☐

15 ÷ 5 = ☐   40 ÷ 5 = ☐   50 ÷ 5 = ☐

2 Write the missing numbers.

☐ ÷ 5 = 2   20 ÷ ☐ = 4   ☐ ÷ 5 = 10

15 ÷ ☐ = 3   ☐ ÷ 5 = 6   10 ÷ ☐ = 2

☐ ÷ 5 = 7   ☐ ÷ 5 = 3   ☐ ÷ 5 = 11

☐ ÷ 5 = 5   50 ÷ ☐ = 10   ☐ ÷ 5 = 9

☐ ÷ 5 = 8   ☐ ÷ 5 = 4   40 ÷ ☐ = 8

Game: Dividing by 5 game

You need: paperclip, pencil and 20 counters

Player 1				Player 2		
60 ÷ 5	35 ÷ 5	20 ÷ 5		25 ÷ 5	15 ÷ 5	60 ÷ 5
45 ÷ 5		50 ÷ 5		30 ÷ 5		35 ÷ 5
25 ÷ 5		55 ÷ 5		50 ÷ 5		40 ÷ 5
15 ÷ 5	30 ÷ 5	40 ÷ 5		55 ÷ 5	45 ÷ 5	20 ÷ 5

- Before you start decide who is Player 1 and who is Player 2.
- Take turns to:
  – Spin the spinner and say the number.
  – Look for a 5 times table division fact that equals that number.
  – Place a counter on that fact on your chart above. If the fact already has a counter on it, miss a turn.
- The winner is the first player to cover 8 of their facts with a counter.

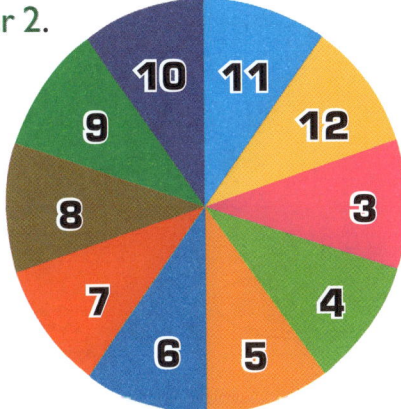

3. Each number that goes into a function machine is divided by 5. Write the numbers that come out of the machines.

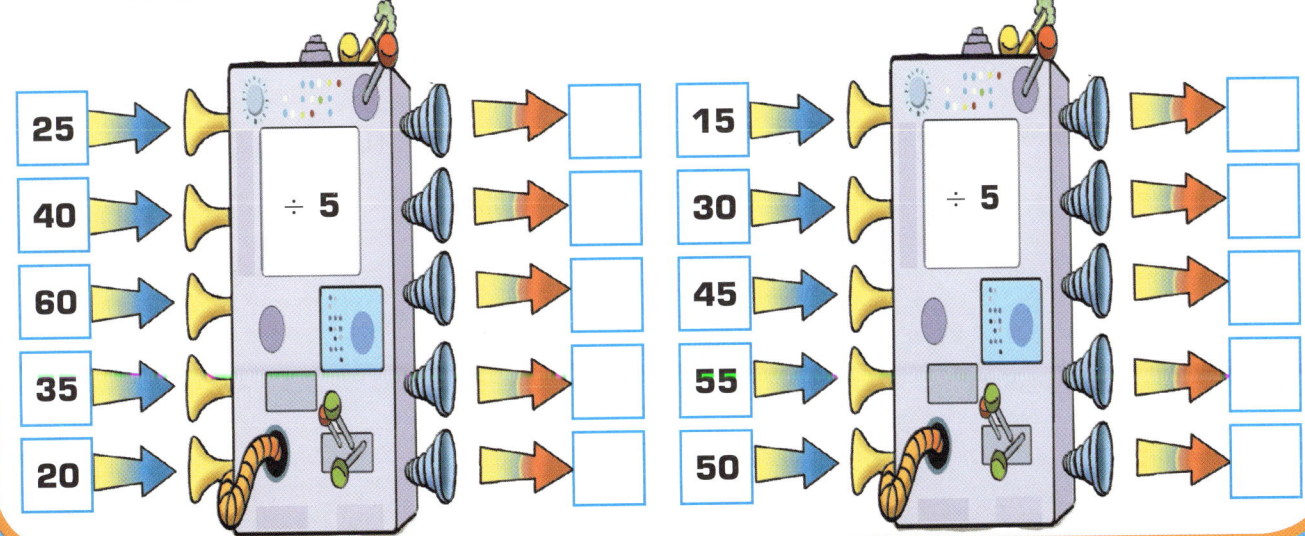

Point to one of the division calculations in the game and ask your child to say the answer. Encourage your child to give quick responses to each question.

10 times table

1 Answer these.

10 × 3 = ☐   8 × 10 = ☐   10 × 7 = ☐

10 × 9 = ☐   6 × 10 = ☐   10 × 12 = ☐

7 × 10 = ☐   10 × 6 = ☐   5 × 10 = ☐

4 × 10 = ☐   10 × 10 = ☐   9 × 10 = ☐

10 × 5 = ☐   11 × 10 = ☐   2 × 10 = ☐

2 Write the missing numbers.

☐ × 10 = 40   ☐ × 10 = 60   10 × ☐ = 80

☐ × 10 = 20   10 × ☐ = 100   10 × ☐ = 40

9 × ☐ = 90   ☐ × 10 = 70   ☐ × 10 = 90

☐ × 10 = 100   ☐ × 10 = 30   ☐ × 10 = 50

10 × ☐ = 80   10 × ☐ = 120   3 × ☐ = 30

Game: 10 times table game

You need: paperclip, pencil and 20 counters

Player 1

| 50 | 90 | 110 | 70 | 120 | 60 | 30 | 100 | 80 | 40 |

Player 2

| 40 | 110 | 100 | 50 | 80 | 120 | 60 | 90 | 30 | 70 |

- Before you start decide who is Player 1 and who is Player 2.
- Take turns to:
  – Spin the spinner and say the number.
  – Multiply the number by 10 and say the answer.
  – Place a counter on that number on your line above. If the number already has a counter on it, miss a turn.
- The winner is the first player to cover 8 of their numbers with a counter.

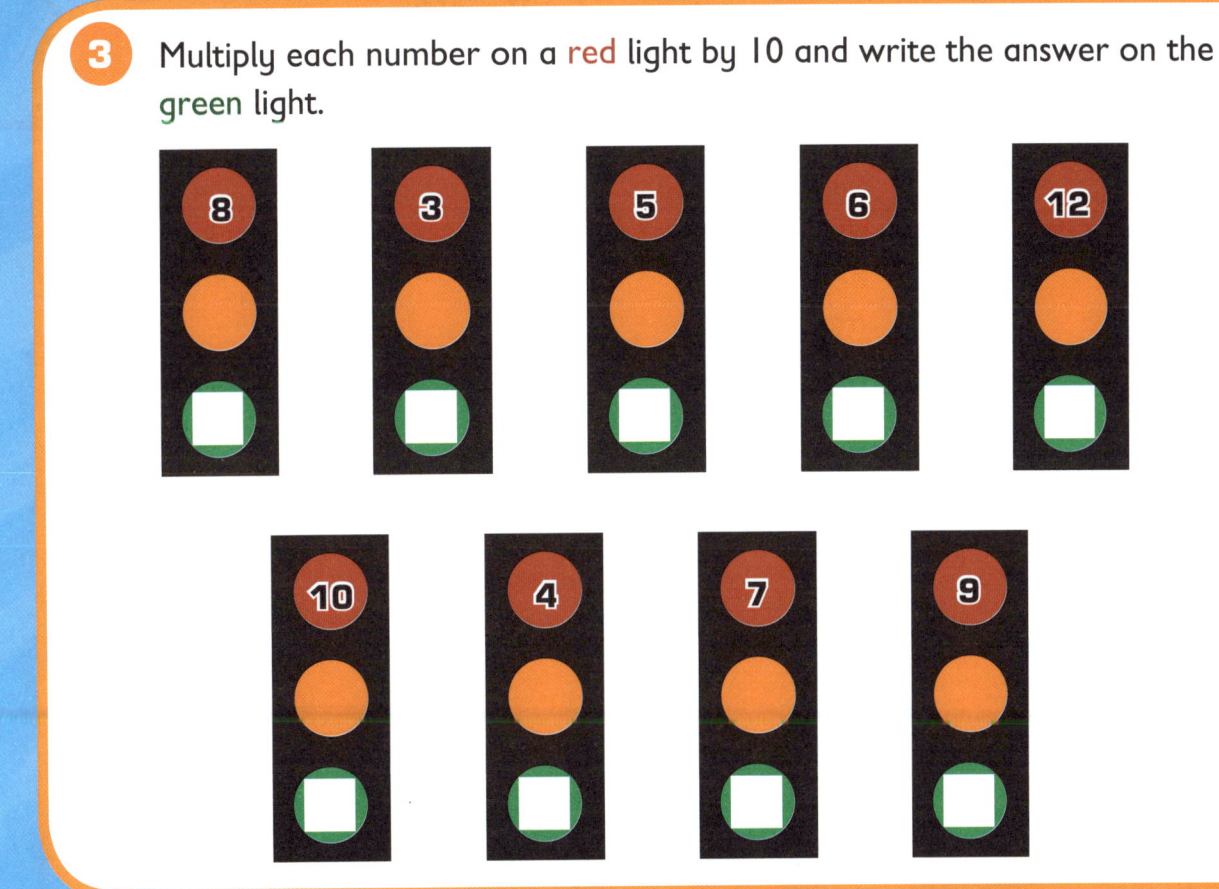

3. Multiply each number on a red light by 10 and write the answer on the green light.

8   3   5   6   12

10   4   7   9

Point to one of the numbers in the game, e.g. 60, and ask your child to tell you the 10 times table fact that has that answer, i.e. 6 × 10.

Division facts related to the 10 times table

1 Answer these.

70 ÷ 10 = ☐   90 ÷ 10 = ☐   60 ÷ 10 = ☐

40 ÷ 10 = ☐   20 ÷ 10 = ☐   80 ÷ 10 = ☐

30 ÷ 10 = ☐   50 ÷ 10 = ☐   110 ÷ 10 = ☐

2 Write the missing numbers.

☐ ÷ 10 = 4   70 ÷ ☐ = 7   ☐ ÷ 10 = 11

30 ÷ ☐ = 3   ☐ ÷ 10 = 6   100 ÷ ☐ = 10

☐ ÷ 10 = 7   ☐ ÷ 10 = 12   ☐ ÷ 10 = 3

☐ ÷ 10 = 2   60 ÷ ☐ = 6   ☐ ÷ 10 = 9

☐ ÷ 10 = 8   ☐ ÷ 10 = 5   80 ÷ ☐ = 8

Game: Dividing by 10 game

You need: paperclip, pencil and 20 counters

Player 1			Player 2		
110 ÷ 10	40 ÷ 10	100 ÷ 10	90 ÷ 10	100 ÷ 10	120 ÷ 10
80 ÷ 10		50 ÷ 10	40 ÷ 10		70 ÷ 10
70 ÷ 10		60 ÷ 10	30 ÷ 10		110 ÷ 10
120 ÷ 10	90 ÷ 10	30 ÷ 10	80 ÷ 10	60 ÷ 10	50 ÷ 10

- Before you start decide who is Player 1 and who is Player 2.
- Take turns to:
  – Spin the spinner and say the number.
  – Look for a 10 times table division fact that equals that number.
  – Place a counter on that fact on your chart above. If the fact already has a counter on it, miss a turn.
- The winner is the first player to cover 8 of their facts with a counter.

3 Divide each number on a red light by 10 and write the answer on the green light.

Point to one of the division calculations in the game and ask your child to say the answer. Encourage your child to give quick responses to each question.

Linking multiplication and division

If you know one times table or division fact then you know three more related times table or division facts.

So, you can use your known fact to help work out the answers to unknown facts.

Example

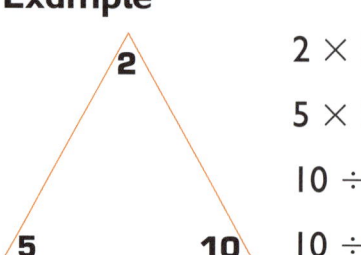

$2 \times 5 = 10$
$5 \times 2 = 10$
$10 \div 2 = 5$
$10 \div 5 = 2$

1 Write the times table and related division facts for each of these.

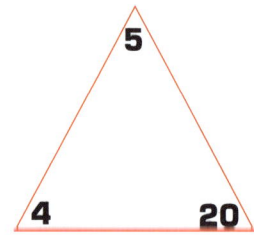

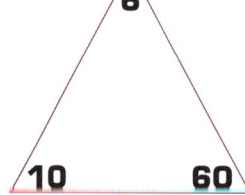

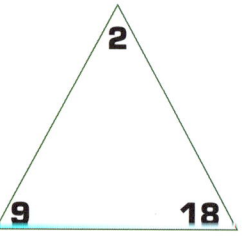

$5 \times \square = \square$

$\square \times 5 = \square$

$\square \div 5 = \square$

$\square \div \square = 5$

$\square \times 10 = \square$

$10 \times \square = \square$

$\square \div \square = 10$

$\square \div 10 = \square$

$2 \times \square = \square$

$\square \times 2 = \square$

$\square \div \square = 2$

$\square \div 2 = \square$

2 For each given number fact, write the three related facts.

$3 \times 2 = 6$

$\square \times \square = \square$

$\square \div \square = \square$

$\square \div \square = \square$

$7 \times 5 = 35$

$\square \times \square = \square$

$\square \div \square = \square$

$\square \div \square = \square$

$120 \div 10 = 12$

$\square \div \square = \square$

$\square \times \square = \square$

$\square \times \square = \square$

30

3 Use the numbers in each triangle to make two times tables and two related division number sentences.

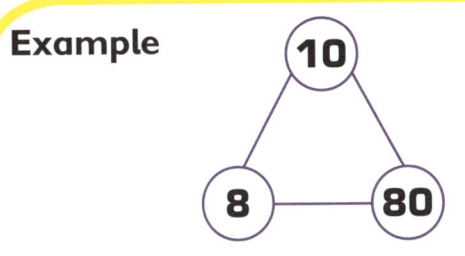

Example

8 × 10 = 80   10 × 8 = 80
80 ÷ 8 = 10   80 ÷ 10 = 8

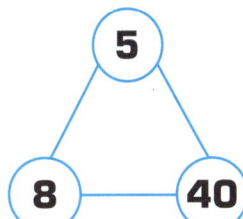
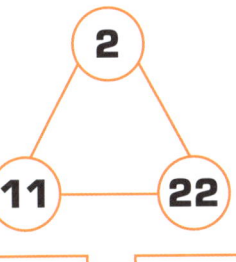
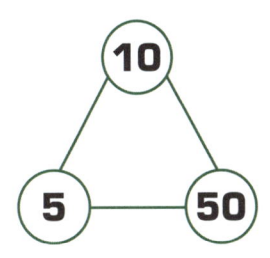

4 Use each set of three numbers to write two times tables and two related division number sentences.

2, 6, 12   10, 9, 90   5, 12, 60

It is important that your child knows by heart the 2, 5 and 10 times tables and the related division facts as they will help when they have to move on to knowing all the multiplication tables up to 12 × 12. Counting on and back in steps of 2, 5 and 10 also helps your child to remember the multiples of 2, 5 and 10.

31

Answers

Understanding multiplication
Page 4

1   $2 + 2 + 2 = 6$ $5 \times 4 = 20$
   $3 \times 2 = 6$
   $4 + 4 + 4 + 4 + 4 = 20$
   $4 \times 6 = 24$ $6 + 6 + 6 + 6 = 24$

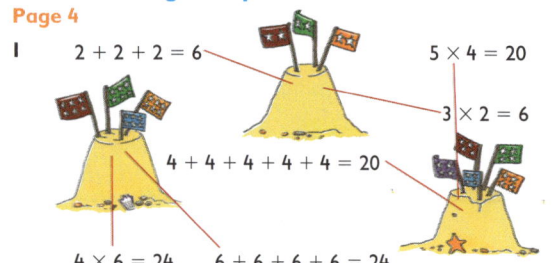

2   $3 + 3 + 3 + 3 + 3 = 15$ $5 \times 3 = 15$
    $2 + 2 + 2 + 2 + 2 + 2 = 12$ $6 \times 2 = 12$
    $5 + 5 = 10$ $2 \times 5 = 10$
    $4 + 4 + 4 + 4 = 16$ $4 \times 4 = 16$

Page 5

3   $2 \times 4 = 8$ $6 \times 3 = 18$ $2 \times 7 = 14$
    $4 \times 2 = 8$
    $3 \times 6 = 18$ $7 \times 2 = 14$

4   $2 \times 8 = 16$ $8 \times 2 = 16$
    $5 \times 6 = 30$ $6 \times 5 = 30$
    $3 \times 4 = 12$ $4 \times 3 = 12$
    $3 \times 7 = 21$ $7 \times 3 = 21$

Understanding division
Page 6

1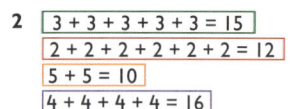
    $10 \div 5 = 2$ $6 \div 2 = 3$ $16 \div 4 = 4$

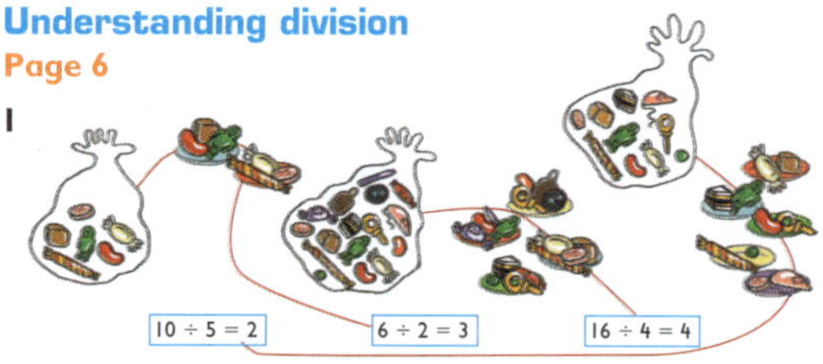

2   $8 \div 4 = 2$ $10 \div 2 = 5$ $9 \div 3 = 3$ $12 \div 6 = 2$

Page 7

3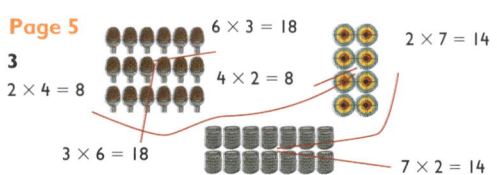
    $6 \div 3 = 2$ $20 \div 5 = 4$ $12 \div 4 = 3$

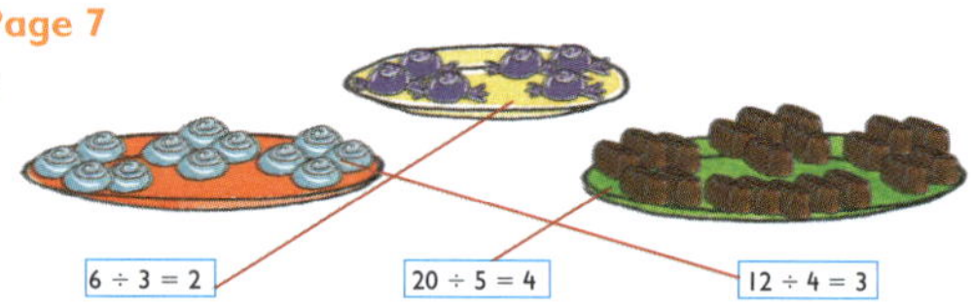

4   $8 \div 2 = 4$ $14 \div 7 = 2$ $25 \div 5 = 5$ $24 \div 4 = 6$

Combining groups of 2, 5 and 10
Page 8

1   $4 \times 2 = 8$ $7 \times 2 = 14$ $3 \times 2 = 6$ $5 \times 2 = 10$
2   $4 \times 5 = 20$ $6 \times 5 = 30$
    $7 \times 5 = 35$ $5 \times 5 = 25$
    $8 \times 5 = 40$ $10 \times 5 = 50$

Page 9

3   $3 \times 10 = 30$ $6 \times 10 = 60$ $2 \times 10 = 20$
    $4 \times 10 = 40$ $7 \times 10 = 70$ $5 \times 10 = 50$
4   $6 \times 2 = 12$ $3 \times 5 = 15$ $8 \times 10 = 80$
    $9 \times 2 = 18$ $12 \times 5 = 60$ $11 \times 10 = 110$

Arranging into equal groups of 2, 5 and 10
Page 10

1   $12 \div 2 = 6$ 2 $20 \div 5 = 4$ $40 \div 5 = 8$
    $10 \div 2 = 5$ $16 \div 2 = 8$ $15 \div 5 = 3$ $25 \div 5 = 5$

Page 11

3   $30 \div 10 = 3$ $50 \div 10 = 5$ 4 $14 \div 2 = 7$ $10 \div 5 = 2$
    $60 \div 10 = 6$ $40 \div 10 = 4$ $35 \div 5 = 7$ $20 \div 10 = 2$

Practising Division
Page 12

1   $8 \div 2 = 4$ 2 $10 \div 5 = 2$
    $14 \div 2 = 7$ $15 \div 5 = 3$
    $10 \div 2 = 5$ $25 \div 5 = 5$
    $18 \div 2 = 9$ $30 \div 5 = 6$

Page 13

3   $20 \div 10 = 2$
    $30 \div 10 = 3$
    $40 \div 10 = 4$
4   $12 \div 3 = 4$ $24 \div 4 = 6$
    $24 \div 6 = 4$ $28 \div 7 = 4$

Counting in steps of 1, 2, 5 and 10
Page 14

1   8, 9, 10, 11, 12, 13, 14, 15   2 25, 30, 35, 40, 45, 50, 55, 60
    5, 10, 15, 20, 25, 30, 35, 40   18, 20, 22, 24, 26, 28, 30, 32
    4, 6, 8, 10, 12, 14, 16, 18   23, 24, 25, 26, 27, 28, 29, 30
    10, 20, 30, 40, 50, 60, 70, 80   30, 40, 50, 60, 70, 80, 90, 100
    15, 16, 17, 18, 19, 20, 21, 22   43, 45, 47, 49, 51, 53, 55, 57

Page 15

3   75, 70, 65, 60, 55, 50, 45, 40   4 34, 33, 32, 31, 30, 29, 28, 27
    90, 80, 70, 60, 50, 40, 30, 20   95, 90, 85, 80, 75, 70, 65, 60
    28, 26, 24, 22, 20, 18, 16, 14   100, 90, 80, 70, 60, 50, 40, 30
    45, 44, 43, 42, 41, 40, 39, 38   65, 60, 55, 50, 45, 40, 35, 30
    37, 35, 33, 31, 29, 27, 25, 23   85, 84, 83, 82, 81, 80, 79, 78

Doubling and halving
Page 16

1   $10 \rightarrow 20$ $2 \rightarrow 4$ $6 \rightarrow 12$ 2 $8 \rightarrow 4$ $6 \rightarrow 3$ $4 \rightarrow 2$
    $5 \rightarrow 10$ $7 \rightarrow 14$ $20 \rightarrow 40$ $2 \rightarrow 1$ $16 \rightarrow 8$ $14 \rightarrow 7$
    $12 \rightarrow 24$ $4 \rightarrow 8$ $14 \rightarrow 28$ $12 \rightarrow 6$ $10 \rightarrow 5$ $18 \rightarrow 9$

2 times table
Page 18

1   $4 \times 2 = 8$ $10 \times 2 = 20$ $9 \times 2 = 18$
    $2 \times 5 = 10$ $3 \times 2 = 6$ $2 \times 8 = 16$
    $7 \times 2 = 14$ $2 \times 6 = 12$ $2 \times 10 = 20$
    $2 \times 9 = 18$ $2 \times 2 = 4$ $11 \times 2 = 22$
    $2 \times 3 = 6$ $8 \times 2 = 16$ $2 \times 7 = 14$
2   $6 \times 2 = 12$ $1 \times 2 = 2$ $2 \times 2 = 4$
    $7 \times 2 = 14$ $8 \times 2 = 16$ $2 \times 10 = 20$
    $9 \times 2 = 18$ $8 \times 2 = 16$ $3 \times 2 = 6$
    $2 \times 12 = 24$ $4 \times 2 = 8$ $9 \times 2 = 18$
    $2 \times 7 = 14$ $2 \times 10 = 20$ $3 \times 2 = 6$

Page 19

3   12   16   8
    14   10   18

32